Words to Know

flippers

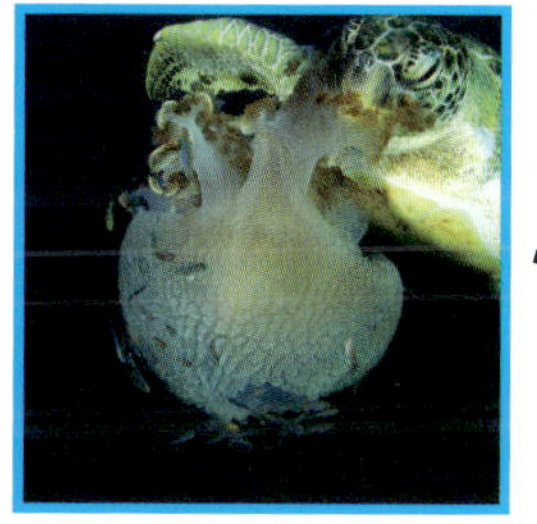

jellyfish

sea turtle

seagrass

shell

This is a **sea turtle**.

UNDER THE SEA ANIMALS

SEA TURTLES

Amy Culliford

TABLE OF CONTENTS

A Pelican Book

Teaching Tips for Caregivers and Teachers:

Research shows that one of the best ways for students to learn a new topic is to read about it.

Before Reading

- Read the title and predict what the book will be about.
- Read the "Words to Know" and discuss the meaning of each word.
- Read the back cover to see what the book is about.

During Reading

- When a student gets to a word that is unknown, ask them to look at the rest of the sentence to find clues to help with the meaning of the unknown word.
- Motivate students with praise and encouragement.

After Reading

- Discuss the main idea of the book.
- Ask students to give one detail that they learned in the book.

Sight Words

a
all
are
eat
four
green
have
is
some
this

sea turtle

Some sea turtles are green.

All sea turtles have four **flippers**.

flippers

All sea turtles have a **shell**.

shell

Some sea turtles eat **seagrass**.

Some sea turtles eat **jellyfish**!

jellyfish

Index

Written by: Amy Culliford
Design by: Under the Oaks Media
Series Development: James Earley
Editor: Kim Thompson

Photos: Willyam Bradberry: cover; Ed Jenkins: p. 4-5; Isabelle Kuehn: p. 7; Richard Whitcombe: p. 9; Salted Life: p. 11; blue-sea: p. 12-13; Rich Carey: p. 15

Library of Congress PCN Data
Sea Turtles / Amy Culliford
Under the Sea Animals
ISBN 978-1-63897-069-9 (hard cover)
ISBN 978-1-63897-155-9 (paperback)
ISBN 978-1-63897-241-9 (EPUB)
ISBN 978-1-63897-327-0 (eBook)
Library of Congress Control Number: 2021945241

Printed in the United States of America.

Seahorse Publishing Company
www.seahorsepub.com

Published in the United States
Seahorse Publishing
PO Box 771325
Coral Springs, FL 33077